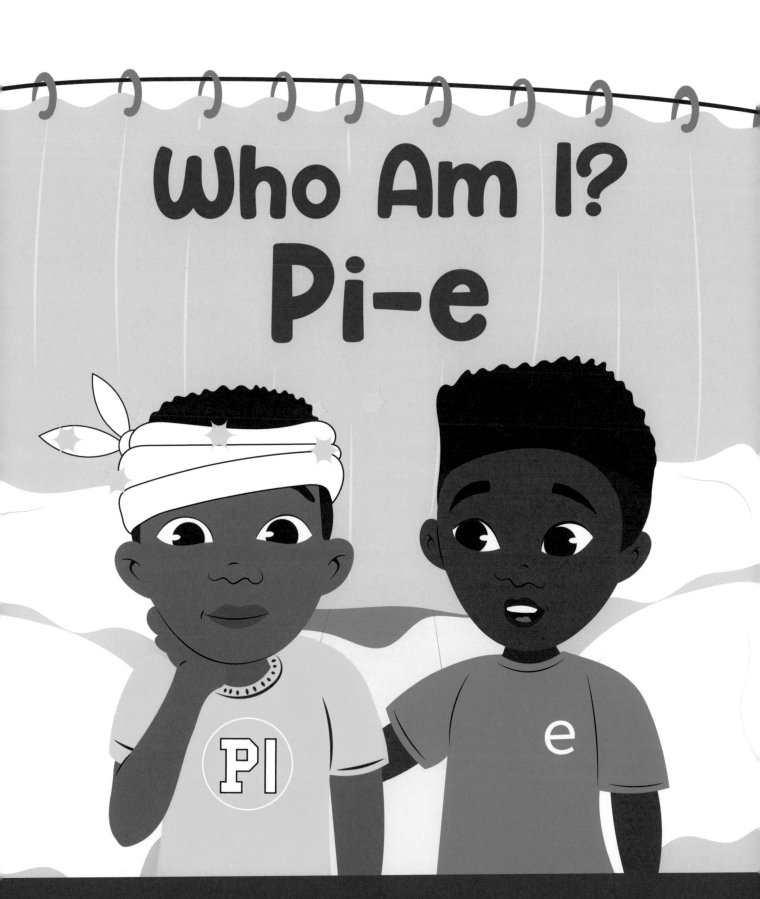

Who Am I?
Pi-e

PANDORA ALEXANDER WALKER

AuthorHouse™
1663 Liberty Drive
Bloomington, IN 47403
www.authorhouse.com
Phone: 833-262-8899

Because of the dynamic nature of the Internet, any web addresses or links contained in this book may have changed since publication and may no longer be valid. The views expressed in this work are solely those of the author and do not necessarily reflect the views of the publisher, and the publisher hereby disclaims any responsibility for them.

Any people depicted in stock imagery provided by Getty Images are models, and such images are being used for illustrative purposes only. Certain stock imagery © Getty Images.

This book is printed on acid-free paper.

ISBN: 979-8-8230-0169-4 (sc)
ISBN: 979-8-8230-0170-0 (e)

Print information available on the last page.

Published by AuthorHouse 02/23/2023

authorHOUSE®

There was a guy named Pi he likes to ride his bike with a friend named Euler, or "e." Pi fell off his bike, hit his head, now he has amnesia. He could not remember just who he was, but Euler knows him quite well and can tell us some facts about Pi.

First thing e says is Pi is smart, friendly, and flexible. This is how I know; pi will change when you need him to be a decimal or fraction even though he is an ir-ra-tion-al number. Irrational numbers do not terminate, cannot be squared, or written as ratios.

Pi and e are unstoppable. Why? They never terminate and continue into infinity.

Next e says pi you have number friends' imaginary number i, 0, 1, summation $\sum$ and infinity ∞ which are not all numbers, but that a conversation for later. So, let's go to practice.

Pi's birthday is March 14ᵗʰ (3.14) Pi and e are numbers used from ancient times until today. Euler's birthday is February 7th (2.7).

The number pi or symbol Π is a <u>mathematical constant</u>, the <u>ratio</u> of a <u>circle's circumference</u> to its <u>diameter</u>, commonly approximated as ≈3.14. Remember pi is flexible. Pi's friend e is ≈2.716 he is flexible too!

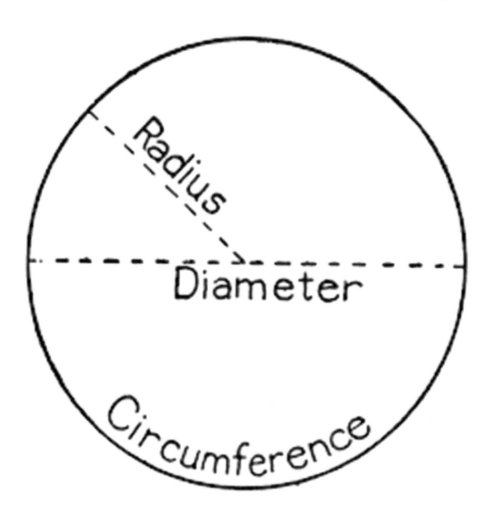

Pi is a non-terminating decimal number 3.14 or 22/7, the number used in most math calculation for area of circle

$$Area\ A = (3.14)r^2$$

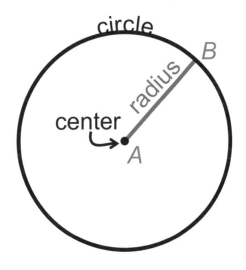

$$Volume\ of\ sphere\quad v = \left(\frac{4}{3}\right)\left(\frac{22}{7}\right)r^3$$

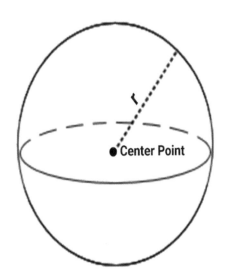

Pi and e are transcendental numbers which means they are not countable like 1, 2, 3....so it cannot be squared like the 1^2, 2^2, or 3^2 so, a circle and square with the same dimensions will not have the same area.

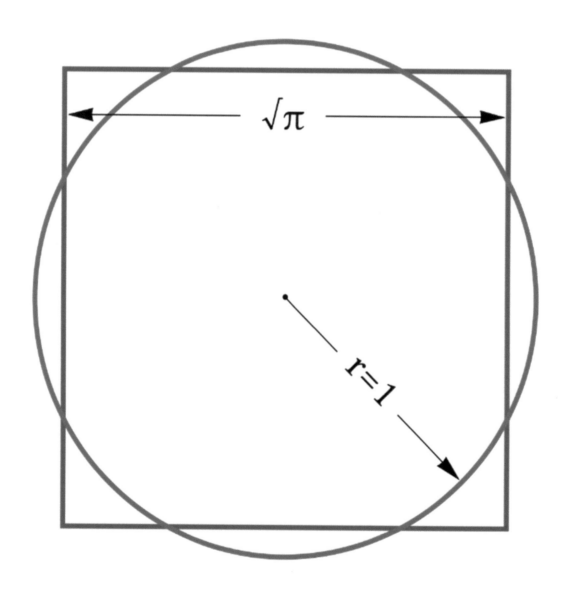

$\sqrt{\pi}$

$r=1$

But square pizza means more pizza and Pi shares with his friends 0, 1, i, and $\sum$ and ∞

Pi and e like pizza but that is not the kind of pie in math for circles, or pie charts which are used to describe and show key facts about food and nutrition using circles. Pi is an irrational number used to calculate area, volume, and circumference.

Pi and e's favorite sport uses statistic a new rating called the player impact estimate, or PIE, which calculates a player's impact on each individual game they play. But does not use pi in the calculation.

Pi's value relates to the circles, spheres, and ellipses, and can be found in everyday life around the house like a fishbowl, bike wheel, or basketball. All circles are ellipses but not all ellipses are circles.

Pi visits the doctor for a checkup and the doctor says he has recovered his memory and his heart checks out too. The human heart follows a normal circular rhythm.

You can find Pi and his friends e, i, $\sum$, ∞ in medicine, art, science, biology, space, statistics, technology, or engineering.

Pi feels much better since e helped him remember who he is, a trans-cen-dental, irrational, non-terminating decimal, sometimes a decimal or fraction when needed. Otherwise, Pi and e will just keep going, going, and going into infinity.

Books Authored by Pandora Alexander Walker

"What's Bugging you about Math? It may
be a simple misunderstanding."
"Queen Bee Mathematical and the Number Garden Friends,"
"The Numbers LOL,"
"Who am I, Pi-e?"
"The Summation of all Things."
"Ooops the Numbers are broken!"
"Queen Bee-tle Mathematical," and "Blessed"

About the Author

Pandora Alexander Walker has a Master of Educational Leadership for grades EC-12, a Bachelor of Science Degree in Mathematics. She has tutored and taught in high school mathematics for over 35 years. She authored this book and eight other books for entertainment and enjoyment.

Acknowledgements

I would like to thank my family, friends, and supporters. A special thanks to Patrice Ellen Ladson Alexander whose time and love went along with us in the journey of Pi-e and all the books in this series.

Dedication

I dedicate this child book to my children and grandsons, Michael Keith, Mariah Ann, Kayson Michael, and Kylan Keith Fontenett

Glossary

Numbers are mathematical values used for counting and measuring objects, and for performing arithmetic calculations. Numbers have various categories like natural numbers, whole numbers, rational and irrational numbers, and so on.

Natural numbers are all positive **numbers** like 1, 2, 3, 4, and so on. They are the **numbers** you usually count, and they continue till infinity.

Example: Natural Numbers or Counting Numbers but **without the zero 1, 2, 3, 4, 5, ...**

Whole numbers are all-natural **numbers** including zero, for example, 0, 1, 2, 3, 4, and so on. Integers include all whole **numbers** and their negative counterpart.

Example: Whole Numbers and natural numbers that include "0" 0, 1, 2, 3, 4, 5, ... (and so on)

Rational Number are numbers that can be written as a Ratio (fraction) of two integers

Example: 1.5 is rational, because it can be written as the ratio 3/2

Irrational Numbers a number that cannot be written as a ratio of two integers ...

Example: π (Pi) is a famous irrational number. π 3.14159265358979323846264433832795... (and more) We cannot write down a simple fraction that equals Pi. The popular approximation of 22/7 3.1428571428571... is close but not accurate.

Transcendental number is a number that is not algebraic or non-terminating decimal.

Example: π (Pi) is a famous irrational number. π 3.14159265358979323846264433832795... Euler's e 2.718281828459045...

Real Numbers are all the above whole, rational and irrational numbers.

Example: Whole Numbers (like 0, 1, 2, 3, ...) Rational Numbers (like 3/4, 0.125, 0.333..., 1.1,) and Irrational Numbers (like π, Euler's "e," $\sqrt{2}$, ...) Real Numbers can also be positive, negative or zero.

Integers are like whole numbers, but they also include *negative* numbers! number line -10 to 10 So, integers can be negative {–1, –2, –3, –4, ...}, positive {1, 2, 3, 4, ...}, or zero {0} We can put that all together like this:

Example: Integers = {..., –4, –3, –2, –1, 0, 1, 2, 3, 4, ...}

Imaginary numbers are numbers that is expressed in terms of the square root of a negative number (usually the square root of –1, represented by i). $i = \sqrt{-1}$

Examples:
$$\sqrt{-9} = \sqrt{9 \times -1} = \sqrt{9} \times \sqrt{-1} = \pm 3i$$
$$\sqrt{-4} = \sqrt{4 \times -1} = \sqrt{4} \times \sqrt{-1} = \pm 2i$$

Summation In mathematics, summation is the addition of a sequence of any kind of numbers, called addends or summands; the result is their sum or total.

Example: summation of [1, 2, 4, 2] is denoted 1 + 2 + 4 + 2, and results in 9,

that is, 1 + 2 + 4 + 2 = 9.

Infinity the concept of something that is unlimited, endless, without bound.

Example: Pi consists of an infinite number of digits. It is often rounded to 3.14 or even 3.14159.... yet no matter how many digits you write, it is impossible to get to the end.

Circle a round plane figure whose boundary (the circumference) consists of points equidistant from a fixed point (the center).

Example:

Sphere a round solid figure, or its surface, with every point on its surface equidistant from its center.

Example:

Ellipse an **oval**. a closed plane curve generated by a point moving in such a way that the sums of its distances from two fixed points is a constant

Example:

P.I.E. Player Impact Estimate aka PIE is a metric to gauge a player's all-around contribution to the game. All statistical categories in the box score engage in the PIE formula. PIE answers what % of the events in a game each player contributed.

Source: Google definition word search various

Author:

All books were written to present basic math facts early to all children for transitional conversations, connections to real world mathematics, reference to align standards for mathematics for enjoyment and entertainment.

Texas Standards TEKS RULE §111.2 kindergarten, Adopted 2012

(6) Geometry and measurement. The student applies mathematical process standards to analyze attributes of two-dimensional shapes and three-dimensional solids to develop generalizations about their properties. The student is expected to:

(A) identify two-dimensional shapes, including circles, triangles, rectangles, and squares as special rectangles.

(B) identify three-dimensional solids, including cylinders, cones, spheres, and cubes, in the real world.

(C) identify two-dimensional components of three-dimensional objects.

(D) identify attributes of two-dimensional shapes using informal and formal geometric language interchangeably.

(E) classify and sort a variety of regular and irregular two- and three-dimensional figures regardless of orientation or size; and

(F) create two-dimensional shapes using a variety of materials and drawings.

RULE §111.3 Grade 1, Adopted 2012

(6) Geometry and measurement. The student applies mathematical process standards to analyze attributes of two-dimensional shapes and three-dimensional solids to develop generalizations about their properties. The student is expected to:

(A) classify and sort regular and irregular two-dimensional shapes based on attributes using informal geometric language.

(B) distinguish between attributes that define a two-dimensional or three-dimensional figure and attributes that do not define the shape.

(C) create two-dimensional figures, including circles, triangles, rectangles, and squares, as special rectangles, rhombuses, and hexagons.

(D) identify two-dimensional shapes, including circles, triangles, rectangles, and squares, as special rectangles, rhombuses, and hexagons and describe their attributes using formal geometric language.

(E) identify three-dimensional solids, including spheres, cones, cylinders, rectangular prisms

(Including cubes), and triangular prisms, and describe their attributes using formal geometric

language.

RULE §111.4 Grade 2, Adopted 2012

(6) Geometry and measurement. The student applies mathematical process standards to analyze attributes of two-dimensional shapes and three-dimensional solids to develop generalizations about their properties. The student is expected to:

(A) classify and sort regular and irregular two-dimensional shapes based on attributes using informal geometric language.

(B) distinguish between attributes that define a two-dimensional or three-dimensional figure and attributes that do not define the shape.

(C) create two-dimensional figures, including circles, triangles, rectangles, and squares, as special rectangles, rhombuses, and hexagons.

(D) identify two-dimensional shapes, including circles, triangles, rectangles, and squares, as special rectangles, rhombuses, and hexagons and describe their attributes using formal geometric language.

(E) identify three-dimensional solids, including spheres, cones, cylinders, rectangular prisms

(Including cubes), and triangular prisms, and describe their attributes using formal geometric language.

RULE §111.5 Grade 3, Adopted 2012

(6) Geometry and measurement. The student applies mathematical process standards to analyze attributes of two-dimensional geometric figures to develop generalizations about their properties. The student is expected to:

(A) classify and sort two- and three-dimensional figures, including cones, cylinders, spheres, triangular and rectangular prisms, and cubes, based on attributes using formal geometric language.

(B) use attributes to recognize rhombuses, parallelograms, trapezoids, rectangles, and squares as examples of quadrilaterals and draw examples of quadrilaterals that do not belong to any of these subcategories.

(C) determine the area of rectangles with whole number side lengths in problems using multiplication related to the number of rows times the number of unit squares in each row.

(D) decompose composite figures formed by rectangles into non-overlapping rectangles to determine the area of the original figure using the additive property of area; and

(E) decompose two congruent two-dimensional figures into parts with equal areas and express the area of each part as a unit fraction of the whole and recognize that equal shares of identical wholes need not have the same shape.

(7) Geometry and measurement. The student applies mathematical process standards to select appropriate units, strategies, and tools to solve problems involving customary and metric measurement. The student is expected to:

(A) represent fractions of halves, fourths, and eighths as distances from zero on a number line.

RULE §111.7 Grade 5, Adopted 2012

(5) Geometry and measurement. The student applies mathematical process standards to classify two-dimensional figures by attributes and properties. The student is expected to classify two-dimensional figures in a hierarchy of sets and subsets using graphic organizers based on their attributes and properties.

(6) Geometry and measurement. The student applies mathematical process standards to understand, recognize, and quantify volume. The student is expected to:

(A) recognize a cube with side length of one unit as a unit cube having one cubic unit of volume and

the volume of a three-dimensional figure as the number of unit cubes (n cubic units) needed to fill it

with no gaps or overlaps if possible.

RULE §111.26 Grade 6, Adopted 2012

(2) Number and operations. The student applies mathematical process standards to represent and use

rational numbers in a variety of forms. The student is expected to:

(A) classify whole numbers, integers, and rational numbers using a visual representation such as a Venn diagram to describe relationships between sets of numbers.

(B) identify a number, its opposite, and its absolute value.

(C) locate, compare, and order integers and rational numbers using a number line.

(D) order a set of rational numbers arising from mathematical and real-world contexts; and

(E) extend representations for division to include fraction notation such as a/b represents the same number as a ÷ b where b ≠ 0.

RULE §111.27 Grade 6, Adopted 2012

(5) Proportionality. The student applies mathematical process standards to use geometry to describe or solve problems involving proportional relationships. The student is expected to:

(A) generalize the critical attributes of similarity, including ratios within and between similar shapes.

(B) describe π as the ratio of the circumference of a circle to its diameter; and

(C) solve mathematical and real-world problems involving similar shape and scale drawings.

RULE §111.28 Grade 8 Adopted 2012

(6) Expressions, equations, and relationships. The student applies mathematical process standards to develop mathematical relationships and make connections to geometric formulas. The student is expected to:

(A) describe the volume formula V = Bh of a cylinder in terms of its base area and its height.

(B) model the relationship between the volume of a cylinder and a cone having both congruent bases and heights and connect that relationship to the formulas; and

(C) use models and diagrams to explain the Pythagorean theorem.

(7) Expressions, equations, and relationships. The student applies mathematical process standards to use geometry to solve problems.

(A) solve problems involving the volume of cylinders, cones, and spheres.

RULE §111.41   HS Geometry, Adopted 2012 (One Credit)

(10) Two-dimensional and three-dimensional figures. The student uses the process skills to recognize characteristics and dimensional changes of two- and three-dimensional figures. The student is expected to:

(A) identify the shapes of two-dimensional cross-sections of prisms, pyramids, cylinders, cones, and spheres and identify three-dimensional objects generated by rotations of two-dimensional shapes; and

(B) determine and describe how changes in the linear dimensions of a shape affect its perimeter, area, surface area, or volume, including proportional and non-proportional dimensional change.

(11) Two-dimensional and three-dimensional figures. The student uses the process skills in the application of formulas to determine measures of two- and three-dimensional figures.

Source: https://texreg.sos.state.tx.us/public/readtac$ext.ViewTAC?tac_view=4&ti=19&pt=2&ch=111

Common Core Standards

Identify and describe shapes.
CCSS.MATH.CONTENT.K.G.A.1
Describe objects in the environment using names of shapes and describe the relative positions of these objects using terms such as *above, below, besides, in front of, behind,* and *next to.*

CCSS.MATH.CONTENT.K.G.A.2
Correctly name shapes regardless of their orientations or overall size.

CCSS.MATH.CONTENT.K.G.A.3
Identify shapes as two-dimensional (lying in a plane, "flat") or three-dimensional ("solid").

Analyze, compare, create, and compose shapes.
CCSS.MATH.CONTENT.K.G.B.4
Analyze and compare two- and three-dimensional shapes, in different sizes and orientations, using informal language to describe their similarities, differences, parts (e.g., number of sides and vertices/"corners") and other attributes (e.g., having sides of equal length).

CCSS.MATH.CONTENT.K.G.B.5
Model shapes in the world by building shapes from components (e.g., sticks and clay balls) and drawing shapes.

CCSS.MATH.CONTENT.K.G.B.6
Compose simple shapes to form larger shapes. *For example, "Can you join these two triangles with full sides touching to make a rectangle?"*

Solve real-life and mathematical problems involving angle measure, area, surface area, and volume.
CCSS.MATH. CONTENT.7. G.B.4
Know the formulas for the area and circumference of a circle and use them to solve problems; give an informal derivation of the relationship between the circumference and area of a circle.

Solve real-world and mathematical problems involving volume of cylinders, cones, and spheres.
CCSS.MATH. CONTENT.8. G.C.9
Know the formulas for the volumes of cones, cylinders, and spheres and use them to solve real-world and mathematical problems.

Explain volume formulas and use them to solve problems

CCSS.MATH.CONTENT.HSG.GMD. A.1

Give an informal argument for the formulas for the circumference of a circle, area of a circle, volume of a cylinder, pyramid, and cone. Use dissection arguments, Cavalieri's principle, and informal limit arguments.

CCSS.MATH.CONTENT.HSG.GMD. A.2

(+) Give an informal argument using Cavalieri's principle for the formulas for the volume of a sphere and other solid figures.

CCSS.MATH.CONTENT.HSG.GMD. A.3

Use volume formulas for cylinders, pyramids, cones, and spheres to solve problems. *

Visualize relationships between two-dimensional and three-dimensional objects

CCSS.MATH.CONTENT.HSG.GMD. B.4

Identify the shapes of two-dimensional cross-sections of three-dimensional objects, and identify three-dimensional objects generated by rotations of two-dimensional objects.

Source: http://www.corestandards.org/standards-in-your-state/

Printed in the United States
by Baker & Taylor Publisher Services